The Serpent River

by

Don Wilcox

Double 9 BOOKS

The Serpent River

by Don Wilcox

ISBN: 978-93-59324-60-9

Published by

DOUBLE 9 BOOKS
2/13-B, Ansari Road, Daryaganj
New Delhi - 110002
info@double9books.com
www.double9books.com
Tel. 011-40042856

ABOUT THE AUTHOR

Don Wilcox, an enigmatic twentieth-century wordsmith, made an everlasting influence on the field of speculative fiction. Wilcox, who was born in Canada in 1905, began a writing career that would lead to him becoming a prolific and significant personality in the worlds of science fiction and fantasy. Their writing helps individuals connect with and understand one another. His writing captured the essence of a quickly changing world, distinguished by inventive leaps and thought-provoking ideas. Wilcox's writings frequently ventured into unexplored territory of science and the human psyche, challenging conventional storytelling. His ability to combine vivid imagery with in-depth philosophical insights won him a strong following among readers looking for intellectually intriguing stories. Wilcox wrote a plethora of short tales and books during his lifetime, each demonstrating his unlimited creativity. His contributions to pulp periodicals of the day, such as "Amazing Stories" and "Weird Tales," demonstrated his ability in producing stories ranging from dystopian futures to fantasy universes. Don Wilcox's literary legacy continues as a monument to the continuing force of speculative fiction, despite the fact that his name is not as well known today as some of his contemporaries.

CONTENTS

"Split" Campbell and I brought our ship down to a quiet landing on the summit of a mile-wide naked rock, and I turned to the telescope for a closer view of the strange thing we had come to see.

It shone, eighteen or twenty miles away, in the light of the late afternoon sun. It was a long silvery serpent-like **something** that crawled slowly over the planet's surface.

There was no way of guessing how large it was, at this distance. It might have been a rope rolled into shape out of a mountain—or a chain of mountains. It might have been a river of bluish-gray dough that had shaped itself into a great cable. Its diameter? If it had been a hollow tube, cities could have flowed through it upright without bending their skyscrapers. It was, to the eye, an endless rope of cloud oozing along the surface of the land. No, not cloud, for it had the compactness of solid substance.

We could see it at several points among the low foothills. Even from this distance we could guess that it had been moving along its course for centuries. Moving like a sluggish snake. It followed a deep-worn path between the nearer hills and the high jagged mountains on the horizon.

What was it?

"Split" Campbell and I had been sent here to learn the answers. Our sponsor was the well known "EGGWE" (the Earth-Galaxy Good Will Expeditions.) We were under the EGGWE Code. We were the first expedition to this planet, but we had come equipped with two important pieces of advance information. The Keynes-Roy roving cameras (unmanned) had brought back to the Earth choice items of fact about various parts of the universe. From these photos we knew (1) that man lived on this planet, a humanoid closely resembling the humans of the Earth; and (2) that a vast cylindrical "rope" crawled the surface of this land, continuously, endlessly.

We had intentionally landed at what we guessed would be a safe distance from the rope. If it were a living thing, like a serpent, we preferred not to disturb it. If it gave off heat or poisonous gases or deadly vibrations, we meant to keep our distance. If, on the other hand, it proved to be some sort of vegetable—a vine of glacier proportions—or a river of some silvery, creamy substance—we would move in upon it gradually, gathering facts as we progressed. I could depend upon "Split" to record all observable phenomena with the accuracy of split-hairs.

Split was working at the reports like a drudge at this very moment.

I looked up from the telescope, expecting him to be waiting his turn eagerly. I misguessed. He didn't even glance up from his books. Rare young Campbell! Always a man of duty, never a man of impulse!

"Here Campbell, take a look at the 'rope'."

"Before I finish the reports, sir? If I recall our Code, Section Two, Order of Duties upon Landing: A—"

"Forget the Code. Take a look at the rope while the sun's on it.... See it?"

"Yes sir."

"Can you see it's moving? See the little clouds of dust coming up from under its belly?"

"Yes sir. An excellent view, Captain Linden."

"What do you think of it, Split? Ever see a sight like that before?"

"No sir."

"Well, what about it? Any comments?"

Split answered me with an enthusiastic, "By gollies, sir!" Then, with restraint, "It's precisely what I expected from the photographs, sir. Any orders, sir?"

"Relax, Split! That's the order. Relax!"

"Thanks—thanks, Cap!" That was his effort to sound informal, though coming from him it was strained. His training had given him an exaggerated notion of the importance of dignity and discipline.

He was naturally so conscientious it was painful. And to top it all, his scientific habit of thought made him want to stop and weigh his words even when speaking of casual things such as how much sugar he required in his coffee.

Needless to say, I had kidded him unmercifully over these traits. Across the millions of miles of space that we had recently traveled (our first voyage together) I had amused myself at his expense. I had sworn that he would find, in time, that he couldn't even trim his fingernails without calipers, or comb his hair without actually physically splitting the hairs that cropped up in the middle of the part. That was when I had nicknamed him "Split"—and the wide ears that stuck out from his stubble-cut blond hair had glowed with the pink of selfconsciousness. Plainly, he liked the kidding. But if I thought I could rescue him from the weight of dignity and duty, I was mistaken.

Now he had turned the telescope for a view far to the right. He paused.

"What do you see?" I asked.

"I cannot say definitely. The exact scientific classification of the object I am observing would call for more detailed scrutiny—"

"You're seeing some sort of object?"

"Yes sir."

"What sort of object?"

"A living creature, sir—upright, wearing clothes—"

"A **man**?"

"To all appearances, sir—"

"You bounder, give me that telescope!"

CHAPTER 2

If you have explored the weird life of many a planet, as I have, you can appreciate the deep sense of excitement that comes over me when, looking out at a new world for the first time, I see a man-like animal.

Walking upright!

Wearing adornments in the nature of clothing!

I gazed, and my lungs filled with the breath of wonderment. A man! Across millions of miles of space—a man, like the men of the Earth.

Six times before in my life of exploration I had gazed at new realms within the approachable parts of our universe, but never before had the living creatures borne such wonderful resemblance to the human life of our Earth.

A man!

He might have been creeping on all fours.

He might have been skulking like a lesser animal.

He might have been entirely naked.

He was none of these—and at the very first moment of viewing him I felt a kinship toward him. Oh, he was primitive in appearance—but had my ancestors not been the same? Was this not a mirror of my own race a million years or so ago? I sensed that my own stream of life had somehow crossed with his in ages gone by. How? Who can ever know? By what faded charts of the movements through the sky will man ever be able to retrace relationships of forms of life among planets?

"Get ready to go out and meet him, Campbell," I said. "He's a friend."

Split Campbell gave me a look as if to say, Sir, you don't even know what sort of animal he is, actually, much less whether he's friendly or murderous.

"There are some things I can sense on first sight, Campbell. Take my word for it, he's a friend."

"I didn't say anything, sir."

"Good. Don't. Just get ready."

"We're going to go **out**—?"

"Yes," I said. "Orders."

"And meet both of them?" Split was at the telescope.

"Both?" I took the instrument from him. Both! "Well!"

"They seem to be coming out of the ground," Split said. "I see no signs of habitation, but apparently we've landed on top of an underground city—though I hasten to add that this is only an hypothesis."

"One's a male and the other's a female," I said.

"Another hypothesis," said Split.

The late evening sunshine gave us a clear view of our two "friends". They were fully a mile away. Split was certain they had not seen our ship, and to this conclusion I was in agreement. They had apparently come up out of the barren rock hillside to view the sunset. I studied them through the telescope while Split checked over equipment for a hike.

The man's walk was unhurried. He moved thoughtfully, one might guess. His bare chest and legs showed him to be statuesque in mold, cleanly muscled, fine of bone. His skin was almost the color of the cream-colored robe which flowed from his back, whipping lightly in the breeze. He wore a brilliant red sash about his middle, and this was matched by a red headdress that came down over his shoulders as a circular mantle.

The girl stood several yards distant, watching him. This was some sort of ritual, no doubt. He was not concerned with her, but with the setting sun. Its rays were almost horizontal, knifing through a break in the distant mountain skyline. He went through some routine motions, his moving arms highlighted by the lemon-colored light of evening.

The girl approached him. Two other persons appeared from somewhere back of her.... Three.... Four.... Five....

"Where do they come from?" Split had paused in the act of checking equipment to take his turn at the telescope. If he had not done so, I might not have made a discovery. The landscape was **moving**.

The long shadows that I had not noticed through the telescope were a prominent part of the picture I saw through the ship's window when I looked out across the scene with the naked eye. The shadows were moving.

They were tree shadows. They were moving toward the clearing where the crowd gathered. And the reason for their movement was that the trees themselves were moving.

"Notice anything?" I asked Split.

"The crowd is growing. We've certainly landed on top of a city." He gazed. "They're coming from underground."

Looking through the telescope, obviously he didn't catch the view of the moving trees.

"Notice anything else unusual?" I persisted.

"Yes. The females—I'm speaking hypothetically—but they **must** be females—are all wearing puffy white fur ornaments around their elbows. I wonder why?"

"You haven't noticed the trees?"

"The females are quite attractive," said Split.

I forgot about the moving trees, then, and took over the telescope. Mobile trees were not new to me. I had seen similar vegetation on other planets—"sponge-trees"—which possessed a sort of muscular quality. If these were similar, they were no doubt feeding along the surface of the slope below the rocky plateau. The people in the clearing beyond paid no attention to them.

I studied the crowd of people. Only the leader wore the brilliant garb. The others were more scantily clothed. All were handsome of build. The lemon-tinted sunlight glanced off the muscular shoulders of the males and the soft curves of the females.

"Those furry elbow ornaments on the females," I said to Split, "they're for protection. The caves they live in must be narrow, so they pad their elbows."

"Why don't they pad their shoulders? They don't have anything on their shoulders."

"Are you complaining?"

We became fascinated in watching, from the seclusion of our ship. If we were to walk out, or make any sounds, we might have interrupted their meeting. Here they were in their native ritual of sunset, not knowing that people from another world watched. The tall leader must be making a speech. They sat around him in little huddles. He moved his arms in calm, graceful gestures.

"They'd better break it up!" Split said suddenly. "The jungles are moving in on them."

"They're spellbound," I said. "They're used to sponge-trees. Didn't you ever see moving trees?"

Split said sharply, "Those trees are marching! They're an army under cover. Look!"

I saw, then. The whole line of advancing vegetation was camouflage for a sneak attack. And all those natives sitting around in meeting were as innocent as a flock of sitting ducks. Split Campbell's voice was edged with alarm. "Captain! Those worshippers—how can we warn them? Oh-oh! Too late. Look!"

All at once the advancing sponge-trees were tossed back over the heads of the savage band concealed within. They were warriors—fifty or more of them—with painted naked bodies. They dashed forward in a wide semicircle, swinging crude weapons, bent on slaughter.

CHAPTER 3

They were waving short clubs or whips with stones tied to the ends. They charged up the slope, about sixty yards, swinging their weird clubs with a threat of death.

Wild disorder suddenly struck the audience. Campbell and I believed we were about to witness a massacre.

"Captain—**Jim**! You're not going to let this happen!"

Our sympathies had gone to the first groups, the peaceable ones. I had the same impulse as Campbell—to do something—anything! Yet here we sat in our ship, more than half a mile from our thirty-five or forty "friends" in danger.

Our friends were panicked. But they didn't take flight. They didn't duck for the holes in the rocky hilltop. Instead, they rallied and packed themselves around their tall leader. They stood, a defiant wall.

"Can we shoot a ray, Jim?"

I didn't answer. Later I would recall that Split **could** drop his dignity under excitement—his "Captain Linden" and "sir." Just now he wanted any sort of split-second order.

We saw the naked warriors run out in a wide circle. They spun and weaved, they twirled their deadly clubs, they danced grotesquely. They were closing in. Closer and closer. It was all their party.

"Jim, can we shoot?"

"Hit number sixteen, Campbell."

Split touched the number sixteen signal.

The ship's siren wailed out over the land.

You could tell when the sound struck them. The circle of savage ones suddenly fell apart. The dancing broke into the wildest contortions you ever saw. As if they'd been spanked by a wave of electricity. The siren scream must have sounded like an animal cry from an unknown world. The attackers ran for the sponge-trees. The rootless jungle came to life. It jerked and jumped spasmodically down the slope. And our siren kept right on singing.

"Ready for that hike, Campbell? Give me my equipment coat." I got into it. I looked back to the telescope. The tall man of the party had behaved with exceptional calmness. He had turned to stare in our direction from the instant the siren sounded. He could no doubt make out the lines of our silvery ship in the shadows. Slowly, deliberately, he marched over the hilltop toward us.

Most of his party now scampered back to the safety of their hiding places in the ground. But a few—the brave ones, perhaps, or the officials of his group—came with him.

"He needs a stronger guard than that," Campbell grumbled.

Sixteen was still wailing. "Set it for ten minutes and come on," I said. Together we descended from the ship.

We took into our nostrils the tangy air, breathing fiercely, at first. We slogged along over the rock surface feeling our weight to be one-and-a-third times normal. We glanced down the slope apprehensively. We didn't want any footraces. The trees, however, were still retreating. Our siren would sing on for another eight minutes. And in case of further danger, we were equipped with the standard pocket arsenal of special purpose capsule bombs.

Soon we came face to face with the tall, stately old leader in the cream-and-red cloak.

Split and I stood together, close enough to exchange comments against the siren's wail. Fine looking people, we observed. Smooth faces. Like the features of Earth men. These creatures could walk down any main street back home. With a bit of makeup they would pass. "Notice, Captain, they have strange looking eyes." "Very smooth." "It's because they have no eyebrows ... no eye lashes." "Very smooth—handsome—attractive."

Then the siren went off.

The leader stood before me, apparently unafraid. He seemed to be waiting for me to explain my presence. His group of twelve gathered in close.

I had met such situations with ease before. "EGGWE" explorers come equipped. I held out a gift toward the leader. It was a singing medallion attached to a chain. It was disc-shaped, patterned after a large silver coin. It made music at the touch of a button. In clear, dainty bell tones it rang out its one tune, "Trail of Stars."

As it played I held it up for inspection. I placed it around my own neck, then offered it to the leader. I thought he was smiling. He was not overwhelmed by the "magic" of this gadget. He saw it for what it was, a token of friendship. There was a keenness about him that I liked. Yes, he was smiling. He bent his head forward and allowed me to place the gift around his neck.

"Tomboldo," he said, pointing to himself.

Split and I tried to imitate his breathy accents as we repeated aloud, "Tomboldo."

We pointed to ourselves, in turn, and spoke our own names. And then, as the names of the others were pronounced, we tried to memorize each breathy sound that was uttered. I was able to remember four or five of them. One was Gravgak.

Gravgak's piercing eyes caused me to notice him. Suspicious eyes? I did not know these people's expressions well enough to be sure.

Gravgak was a guard, tall and muscular, whose arms and legs were painted with green and black diamond designs.

By motions and words we didn't understand, we inferred that we were invited to accompany the party back home, inside the hill, where we would be safe. I nodded to Campbell. "It's our chance to be guests of Tomboldo." Nothing could have pleased us more. For our big purpose—to understand the Serpent River—would be forwarded greatly if we could learn, through the people, what its meanings were. To analyze the river's substance, estimate its rate, its weight, its temperature, and to map its course—these facts were only a part of the information we sought. The fuller story would be to learn how the inhabitants of this planet regarded it: whether they loved or shunned it, and what legends they may have woven around it. All this knowledge would be useful when future expeditions of men from the Earth followed us (through EGGWE) for an extension of peaceful trade relationships.

Tomboldo depended upon the guard Gravgak to make sure that the way was safe. Gravgak was supposed to keep an eye on the line of floating trees that had taken flight down the hillside. Danger still lurked there, we knew. And now the siren that had frightened off the attack was silent. Our ship, locked against invaders, could be forgotten. We were guests of Tomboldo.

Gravgak was our guard, but he didn't work at it. He was too anxious to hear all the talk. In the excitement of our meeting, everyone ignored the growing darkness, the lurking dangers. Gravgak confronted us with agitated jabbering:

"Wollo—yeeta—vo—vandartch—vandartch! Grr—see—o—see—o—see—o!"

"See—o—see—o—see—o," one of the others echoed.

It began to make sense. They wanted us to repeat the siren noises. The enemy had threatened their lives. There could very well have been a wholesale slaughter. But as long as we could make the "see—o—see—o" we were all safe.

Split and I exchanged glances. He touched his hand to the equipment jacket, to remind me we were armed with something more miraculous than a yowling siren.

"See—o—see—o—see—o!" Others of Tomboldo's party echoed the demand. They must have seen the sponge-trees again moving toward our path. "**See—o—see—o!**"

Our peaceful march turned into a spasm of terror. The sponge-trees came rushing up the slope, as if borne by a sudden gust of wind. They bounced over our path, and the war party spilled out of them.

Shouting. A wild swinging of clubs. And no cat-and-mouse tricks. No deliberate circling and closing in. An outright attack. Naked bodies gleaming in the semi-darkness. Arms swinging weapons, choosing the nearest victims. The luminous rocks on the ends of the clubs flashed. Shouting, screeching, hurling their clubs. The whizzing fury filled the air.

I hurled a capsule bomb. It struck at the base of a bouncing sponge-tree, and blew the thing to bits.

The attackers ran back into a huddle, screaming. Then they came forward, rushing defiantly.

Our muscular guard, Gravgak was too bold. He had picked up one of their clubs and he ran toward their advance, and to all of Tomboldo's party it must have appeared that he was bravely rushing to his death. Yet the gesture of the club he swung so wildly could have been intended as a **warning**! It could have meant, Run back, you fools, or these strange devils will throw fire at you.

I threw fire. And so did my lieutenant. He didn't wait for orders, thank goodness. He knew it was their lives or ours. Zip, zip, zip—BLANG-BLANG-BLANG! The bursts of fire at their feet ripped the rocks. The spray caught them and knocked them back. Three or four warriors in the fore ranks were torn up in the blasts. Others were flattened—and those who were able, ran.

They ran, not waiting for the cover of sponge-trees. Not bothering to pick up their clubs.

But the operation was not a complete success. We had suffered a serious casualty. The guard Gravgak. He had rushed out too far, and the first blast of fire and rock had knocked him down. Now Tomboldo and others of the party hovered over him.

His eyes opened a little. I thought he was staring at me, drilling me with suspicion. I worked over him with medicines. The crowd around us stood back in an attitude of awe as Split and I applied ready bandages, and held a stimulant to his nostrils that made him breath back to consciousness.

Suddenly he came to life. Lying there on his back, with the club still at his fingertips, he swung up on one elbow. The swift motion caused a cry of joy from the crowd. I heard a little of it—and then blacked out. For as the muscular Gravgak moved, his fingers closed over the handle of the club. It whizzed upward with him—apparently all by accident. The stone that dangled from the end of the club crashed into my head.

I went into instant darkness. Darkness, and a long, long silence.

CHAPTER 4

Vauna, the beautiful daughter of Tomboldo, came into my life during the weeks that I lay unconscious.

I must have talked aloud much during those feverish hours of darkness.

"Campbell!" I would call out of a nightmare. "Campbell, we're about to land. Is everything set? Check the instruments again, Campbell."

"S-s-sh!" The low hush of Split Campbell's voice would somehow penetrate my dream.

The voices about me were soft. My dreams echoed the soft female voices of this new, strange language.

"Campbell, are you there?... Have you forgotten the Code, Campbell?"

"Quiet, Captain."

"Who is it that's swabbing my face? I can't see."

"It's Vauna. She's smiling at you, Captain. Can't you see her?"

"Is this the pretty one we saw through the telescope?"

"One of them."

"And what of the other? There were two together. I remember—"

"Omosla is here too. She's Vauna's attendant. We're all looking after you, Captain Linden. Did you know I performed an operation to relieve the pressure on your brain? You must get well, Captain." The words of Campbell came through insistently.

After a silence that may have lasted for hours or days, I said, "Campbell, you haven't forgot the EGGWE Code?"

"Of course not, Captain."

"Section Four?"

"Section Four," he repeated in a low voice, as if to pacify me and put me to sleep. "Conduct of EGGWE agents toward native inhabitants: A, No agent shall enter into any diplomatic agreement that shall be construed as binding—"

I interrupted. "Clause D?"

He picked it up. "D, no agent shall enter into a marriage contract with any native.... H-m-m. You're not trying to warn me, are you, Captain Linden? Or are you warning **yourself**?"

At that moment my eyes opened a little. Swimming before my blurred vision was the face of Vauna. I did remember her—yes, she must have haunted my dreams, for now my eyes burned in an effort to define her features more clearly. This was indeed Vauna, who had been one of the party of twelve, and had walked beside her father in the face of the attack. Deep within my subconscious the image of her beautiful face and figure had lingered. I murmured a single word of answer to Campbell's question. "Myself."

In the hours that followed, I came to know the soft footsteps of Vauna. The caverns in which she and her father and all these Benzendella people lived were pleasantly warm and fragrant. My misty impressions of their life about me were like the first impressions of a child learning about the world into which he has been born.

Sometimes I would hear Vauna and her attendant Omosla talking together. Often when Campbell would stop in this part of the cavern to inquire about me, Omosla would drop in also. She and Campbell were learning to converse in simple words. And Vauna and I—yes. If I could only avoid blacking out.

I wanted to see her.

So often my eyes would refuse to open. A thousand nightmares. Space ships shooting through meteor swarms. Stars like eyes. Eyes like stars. The eyes of Vauna, the daughter of Tomboldo. The sensitive stroke of Vauna's fingers, brushing my forehead, pressing my hand.

I regained my health gradually.

"Are you quite awake?" Vauna would ask me in her musical Benzendella words. "You speak better today. Your friend Campbell has brought you more recordings of our language, so you can learn to speak more. My father is eager to talk with you. But you must sleep more. You are still weak."

It gave me a weird sensation to awaken in the night, trying to adjust myself to my surroundings. The Benzendellas were sleep-singers. By night they murmured mysterious little songs through their sleep. Strange harmonies whispered through the caves.

And if I stirred restlessly, the footsteps of Vauna might come to me through the darkness. In her sleeping garments she would come to me, faintly visible in the pink light that filtered through from some corridor.

She would whisper melodious Benzendella words and tell me to go back to sleep, and I would drift into the darkness of my endless dreams.

The day came when I awakened to see both Vauna and her father standing before me. Stern old Tomboldo, with his chalk-smooth face and not a hint of an eyebrow or eyelash, rapped his hand against my ribs, shook the fiber bed lightly, and smiled. From a pocket concealed in his flowing cape, he drew forth the musical watch, touched the button, and played, "Trail of Stars."

"I have learned to talk," I said.

"You have had a long sleep."

"I am well again. See, I can almost walk." But as I started to rise, the wave of blackness warned me, and I restrained my ambition. "I will walk soon."

"We will have much to talk about. Your friend has pointed to the stars and told me a strange story of your coming. We have walked around the ship. He has told me how it rides through the sky. I can hardly make myself believe." Tomboldo's eyes cast upward under the strong ridge of forehead where the eyebrows should have been. He was evidently trying to visualize the flight of a space ship. "We will have much to tell each other."

"I hope so," I said. "Campbell and I came to learn about the **serpent river**." I resorted to my own language for the last two words, not knowing the Benzendella equivalent. I made an eel-like motion with my arm. But they didn't understand. And before I could explain, the footsteps of other Benzendellas approached, and presently I looked around to see that quite an audience had gathered. The most prominent figure of the new group was the big muscular guard of the black and green diamond markings—Gravgak.

"You get well?" Gravgak said to me. His eyes drilled me closely.

"I get well," I said.

"The blow on the head," he said, "was not meant."

I looked at him. Everyone was looking at him, and I knew this was meant to be an occasion of apology. But the light of fire in Vauna's eyes told me that she did not believe. He saw her look, and his own eyes flashed darts of defiance. With an abrupt word to me, he wheeled and started off. "Get well!"

The crowd of men and women made way for him. But in the arched doorway he turned. "Vauna. I am ready to speak to you alone."

She started. I reached and barely touched her hand. She stopped. "I will talk with you later, Gravgak."

"Now!" he shouted. "Alone."

He stalked off. A moment later Vauna, after exchanging a word with her father, excused herself from the crowd and followed Gravgak.

From the way those in the room looked, I knew this must be a dramatic moment. It was as if she had acknowledged Gravgak as her master—or her lover. He had called for her. She had followed.

But her old father was still the master. He stepped toward the door. "Vauna!... Gravgak!... Come back."

(I will always wonder what might have happened if he hadn't called them! Was my distrust of Gravgak justified? Had I become merely a jealous lover—or was I right in my hunch that the tall muscular guard was a potential traitor?)

Vauna reappeared at once. I believe she was glad that she had been called back.

Gravgak came sullenly. At the edge of the crowd in the arched doorway he stood scowling.

"While we are together," old Tomboldo said quietly, looking around at the assemblage, "I must tell you the decision of the council. Soon we will move back to the other part of the world."

There were low murmurs of approval through the chamber.

"We will wait a few days," Tomboldo went on, "until our new friend—" he pointed to me—"is well enough to travel. We would never leave him here to the mercy of the savage ones. He and his helper came through the sky in time to save us from being destroyed. We must never forget this kindness. When we ascend the **Kao-Wagwattl**, the ever moving **rope of life**, these friends shall come with us. On the back of the Kao-Wagwattl **they shall ride with us across the land**."

CHAPTER 5

From that moment on, there was more buzzing around the caverns than a hive of bees. It was like a spaceport before the blastoff of a big interplanetary liner. The excitement was enough to cause a sick man to have a relapse—or get well in a hurry to join in on the commotion. I did my best to get well quick!

"Where is Campbell? Bring me my friend Campbell, please."

Omosla, the pretty attendant and companion of Vauna, was always glad, I noticed, to be sent on an errand to Split Campbell, wherever he was.

From all reports he was reinforcing the defenses at one point or another where these caverns led up to the surface. They told me he was a busy man. The attacks of the savage ones had grown more vicious. They had evidently learned that the Benzendellas intended to move back to other lands; so they had grown bold in their raids, attempting to steal not only the Benzendellas' treasurers but also their women. They had not been successful. My good lieutenant, navigator and scientist, equipped with capsule explosives, had blown one group of them into a fountain of dismembered arms and legs. I could just picture him hurling those miniature bombs at the split-second when they would create the most panic.

The Benzendellas had been quick to recognize a good thing. They only wished he were quadruplets or better, to stand guard continuously at many entrances. They brought him their rare foods, and furnished him with a comfortable couch; they offered him gifts. In short, they loved him for his efficiency, and for himself. Especially (according to the rumors that reached my ears) Omosla.

Pretty little Omosla, I fear, loved him with a love that might have overwhelmed a lesser man. But I knew that Split Campbell would not be swerved. He was devoted to duty, dignity, and the Code. The Code forbade intermarriage with the natives.

Why did I keep thinking of the Code? It shouldn't have crossed my thoughts so often. I hardly dared stop to ask myself what continually brought it to mind. But I knew. The flare of jealously I had felt when Gravgak had tried to call Vauna away from the crowd....

"You are feeling better, Captain?" Vauna said to me as she watched me pace the floor. "You find that you can walk, so you keep walking?"

"I need to walk so I can think."

"If you wish to think, you should sit out on the hillside at the time of sunset. You understand my words?"

"I understand," I smiled. Then, rashly, I added, "I understand your words. I don't always understand you."

"And you wish to understand me?"

"Yes."

"Why?"

I could think of more answers than my vocabulary could handle. I said simply, "When I go back to my own world I should be able to say that I understand the people of this world."

"But you **do** understand us. You see how we live. You hear how we talk. There." She pressed my hand. "That is all you need to understand, isn't it? I am the one who does not understand you."

"How do you mean?"

"I do not see how you live. I do not hear how you talk." She gave a little laugh. "Only see how you walk when you think, but I do not know what you think."

"I think about you," I said.

"That is very nice. I think about you, too, Jim. Since the night you saved us from the savage ones, I have thought about you."

I stopped walking in circles and looked at her. The soft light from the luminous rock walls gave an ivory tint to her bare shoulders. She wore a dress of soft woven material, designed with a diagonal line of little hand-painted sponge-trees. From the curve of her breasts to the lithe gracefulness of her thighs, the close-fitting garment accentuated her beauty.

She was backing away from me, smiling as if wondering if I would follow her. Her arms were bare except for the ornaments of fur around her elbows. These were evidently an insignia of Benzendella womanhood, for no woman of this realm was to be seen without them.

"Come," Vauna said, beckoning me. "Put your ear against the wall. What do you hear?"

She pressed her head against the wall and I did the same. Finally I made out the faint vibrations of some distant rumbling. I asked, "What is it?"

"Kao-Wagwattl."

"The round river that moves like a serpent?"

"It is an endless rope," she said. "It is life."

"Life?"

"It gathers water and food within itself. It gives life to those who seek life. It gives life—"

She stopped, and her pretty poetic expression vanished. My hands touched her hands, my fingers moved gently along her wrists, her forearms—then as my touch neared her fur-covered elbows, a look of shock came into her eyes. "Jim!"

"Yes, Vauna?"

"I was trying to tell you—"

"**What?**"

For a moment she only looked at me, searching my eyes. "We **don't** understand each other, do we?"

Finally I said, "Then why don't we ask each other questions?"

"Yes.... Yes, ask me questions."

"All right." I had an impulse to start pacing again. I walked about for a moment. "Tell me, Vauna. When your friend Gravgak demanded that you come and talk with him alone, what would have happened if your father hadn't called you back?"

She smiled faintly. "I will tell you a secret, Jim. I had already made my father promise to call me back. I whispered to him, 'Call me back.'"

"Why?"

She gave an evasive little laugh. "You understand enough already. Now it is my question. Tell me, Captain Jim, why do you keep saying that you are going back to another world?"

"Because I am. That's my duty."

"When you ride with us on the Kao-Wagwattl you will come with us to another part of this world. It is more beautiful than here. We are only a few. Our race lives in the other part. My father came here only to study, but soon the Kao-Wagwattl will take us all back. And you and your friend Campbell will go with us and belong to us."

The self-discipline of an EGGWE agent is supposed to be his defense against any natives' invitations, no matter how beautiful or charming the native. All I could say was, "You don't understand us, do you, Vauna?"

"Don't I?"

"Your people I love. And you, Vauna. But our orders are to return. I must not think of disobeying my orders. And I assure you Campbell is one who would never disobey."

"The big silver shell will take you away from us?"

"Yes."

"You will remember me?"

"Yes, always."

"Thank you, Jim." She was weeping. I started to take her in my arms, but thought better of it. She dried her eyes. "I will remember you too. When I see Campbell and Omosla, I will have a dream of this hour, and how we didn't understand."

I was quick to make a correction. "You'll not be seeing Campbell. I'll have to take him back with me, you know."

"No, he will be here. It is our rule that he should stay."

"Why?"

"Because he has become the mate of our girl, Omosla."

I looked at her, not believing I had heard her words correctly. A fever swept my brain. In my own language I said harshly, "It's a lie! Campbell would never violate—"

"I do not understand your words," Vauna said softly.

Then in my broken Benzendella accents I asserted, calmly but decisively, "I don't believe what you say. I don't believe that Campbell has become the mate of Omosla."

"You will believe," Vauna said, "when Omosla's baby is born."

CHAPTER 6

I had already sent for Campbell. Mentally I chastized myself for having sent Omosla. For if what I had been told was true, his life had become complicated enough already. (I must admit that for the moment I had something less than proper consideration for **her**.)

Omosla didn't return from the errand for Campbell. Maybe the news of my concern for him had frightened her away. One of her friends told me that Campbell was out on the surface somewhere; that he couldn't be located just now. When he returned they would send him to **me**.

I then sought the counsel of Tomboldo.

"It can't be true, this story about Campbell," I said. "There's been some mistake."

Tomboldo's answer was soft spoken. "Much has happened. You have been ill for many weeks. You must take our word. Do you find the news not to your liking? Omosla is a devoted girl. And if our hero Campbell became her husband, all of us would be proud."

There was no use talking of the EGGWE Code to him, that was plain. All I could say at the moment was, "I'll talk with Campbell."

For the next few nights, after the whole cavern city seemed to be asleep, I would walk forth a little distance. This was more than pacing. It was a test of my strength and my wits, and most of all my confidence that I would not black out. It was proof to myself that I was a well man again. It was a willful act of striking out on my own purposes. I would find Campbell.

Each night I ventured a little farther. The artificial lights burned low. All was quiet. The luminous rock walls stared out from among the cavern furnishings. I walked steadily. I was getting used to the planet's stronger gravity. I was learning to like the sandals they had given me to wear, cushioned with shreds of sponge-tree vegetation.

Tonight as always I walked to the right from the arch, through one of Tomboldo's rooms, and on past the storage rooms. The way opened into a long amber-lighted tunnel. The city branched off in little tunneled avenues from this passageway. Would Campbell be found on guard tonight—this way—or this way—or—

I heard light footsteps, sounds of two persons somewhere in the distance. I moved back toward Tomboldo's part of the cave to wait until the ways had cleared.

Two men were coming through the corridor, conversing in low whispers.

I moved back into the shadows, scarcely breathing.

The glow of amber light from the corridor revealed them, silhouetted. The taller man was driving the smaller one ahead of him, threatening him with a short-bladed knife.

They slowed their steps. Their low whispers were audible.

"If you breathe a word I'll rip you." The agitated words of the tall guard, Gravgak. The light revealed the lines of green-and-black diamonds painted on his thighs.

The smaller man, also a guard, muttered, "Have I ever told anything?"

"You understand, then," said Gravgak. "If anything happens, you'll swear there was an intruder—one of the savages."

"I'll swear it. I'll say that I—"

"Say that he knocked you down and forced his way in. Like this!" Gravgak struck him with his fist. The guard tumbled in a heap against the cavern wall. He lay there, eyes closed. Gravgak tiptoed past my hiding place. His eyes glinted with purpose. He paused at Tomboldo's door, weighed the knife in his hand, then sheathed it. He went on toward Vauna's room.

I skipped to one side of the storage room where I had seen my equipment coat hanging. Without it I could have been no match for this man. My fingers caught it off the wall, I got into it as I hurried back. Automatically my hands checked the contents, everything in place—

Gravgak was conversing with Vauna through the partly opened door. "I told you I would come."

"You have no right. I told you—" There was strength, not fear, in Vauna's low voice.

"Your father means for me to win you, if necessary by force."

"You lie. Go or I'll sound the alarm."

"You are in love with that stranger." His voice trembled with rage. "See, you don't answer. If you want him to live, get rid of him. Send him back in his silver shell."

"You threaten my father's guest?"

"The great Tomboldo will not live long. I have heard the savages plan to come in some night soon and murder him."

At that instant old Tomboldo's voice sounded from the next room. "Who's there, Vauna?"

"Gravgak!" It was Gravgak himself who answered. "I came to protect you, Tomboldo. There's danger—"

Tomboldo's voice thundered with anger at this unaccountable intrusion. "What do you mean?"

"They mean to kill you, and if they do—"

"Who?"

"The savages. And if they succeed, I am your successor. Tell your daughter it's so. Tell her that if a knife blade descends from some dark corner—**look out**! Someone behind you!"

It was a ruse to cause old Tomboldo to whirl about and turn his back to Gravgak. Tomboldo didn't whirl. But he must have seen what I saw, glittering in the dim light—the knife in Gravgak's hand. It flashed up—

I flung a capsule bomb at the arch. Fire flashed, and the voices were swallowed up in the concussion.

CHAPTER 7

The swirl of yellow dust sifted through the cavern passages. Coughing and puffing hard, I fought my way into the heap—in time to catch sight of Gravgak staggering off toward an exit tunnel.

The three of us stood together. A strange trio. Two Benzendellas, one Earth man. Bound together in an allegiance that all the space in the universe could never divide. Vauna was weeping softly, holding her arms tight about herself, her hands cupped over the fur wrappings of her elbows.

She said she could not understand Gravgak's behavior. Once he had had a chance to become the leader. Was it all because he was insane with jealousy—because she loved me?

Her father thought it was more than this. He had evidently read signs of disloyalty in Gravgak, even before my coming. Too many plans had filtered out to the savage enemies. For a long time Gravgak had been impatient for a chance to succeed Tomboldo; my coming had thwarted the original plan—the murderous attack on the sunset meeting. Yes, Gravgak had been twisting the sponge-tree bands into his own schemes even then.

The fine boldness showed in Tomboldo's eyes as he talked. People had gathered, and they saw clearly the truth of his charges.

But now there were delays in getting ready to go to the better land on another side of this planet. Part of the delay was caution. Gravgak would probably lie in waiting for the Benzendella migration to the serpent river. He would plan an attack. Some waiting, some scouting and much preparation would be a matter of wisdom. Meanwhile, if Gravgak could be found, let him be killed on sight.

Several weeks passed. Secret preparations for the twenty mile migration were completed. I was pleased to hear that Campbell had had a share in these plans. He had made several night hikes back to the ship, and had kept watch through the telescope by day, and made valuable observations by means of infra-red photography by night. He knew where the nests of the savage bands were located. Moreover, I learned that he and a few of Tomboldo's choice scouts, under cover of darkness, crossed through the sponge-tree area to examine the Serpent River at close range and determine upon a suitable place for getting the Benzendella tribe aboard.

For these observations, and for an abundance of scientific data which he picked up about the Serpent River itself, I was deeply grateful. If this expedition succeeded in its purposes, the success would be to his credit, not mine.

Nevertheless, when I was at last conducted to his quarters at the end of one of the tunnels—my long awaited visit—I did not spend all my time complimenting him for his fine achievements.

"You're going to be ready to make the trip with the tribe, I presume?" I asked, when we got around to the plans for the migration.

"And leave the ship here? I shall follow orders, Captain, but I should prefer to stay with the ship, and to proceed with the remainder of the scientific assignments."

He handed his field glasses over to one of the relief guards, and led me to a bench in his primitive quarters. A slice of sunlight knifed through from the out-of-doors, the first I had seen for a long time.

"A little sunlight's not a bad thing," I said casually. "I've been needing a little light."

He looked up at me as if he knew what was coming. "If you've been hearing a rumor, don't believe it."

"You've heard it too?"

"They say I'm supposed to become the husband of Omosla."

"All I want is your word, Lieutenant Campbell," I said.

"My word. Captain." Split said dryly. "You know I wouldn't break the Code."

"I believe you.... Okay, we're in a spot. The fact is, the girl's going to have a baby. When she does, she'll declare you her mate. And the tribe will be proud. Have you thought this through?"

"I've tried to."

I began to pace. "You know we can't afford to offend the tribe. If you bluntly deny that you've had anything to do with the girl, they'll be insulted. They're ready to believe her, not you."

"How soon will the child be born?"

"Within a few days."

"How long have we been here?"

"Long enough."

"Why doesn't her true mate speak up, whoever he is?"

I said, "That's one of the strange circumstances. I haven't heard them mention any other man but you. You see, Split, you're the hero of the hour. You're the one they want."

"I hope you're not suggesting that **I** marry this girl."

"I **haven't** suggested it, have **I**? But I will ask this: Do you like the girl?... Love her??... Enough to marry her?"

"Under more favorable conditions—yes. I've never loved anybody before. But Omosla—from the first time I saw her, that evening, in the sunset—"

"All right, Split. But you still tell me you haven't made love to her?"

"Absolutely, **no**. You may not know it, Jim, but I was with you almost constantly for days and nights after your knockout. You came through the operation—the riskiest thing I ever tried in my life. When you began to pull out of it, I could have gladly taken you back to the ship and blasted off for home. But they were giving you care—Vauna and Omosla—and damned intelligent care, according to my orders. By that time the savages were knocking on our doors again, and I went onto the defense job with my pockets full of scare bombs, and the other kind too. From then on, I couldn't have held to tighter discipline if I'd been in a planetary war, I swear it."

I beat my fist lightly on Split's shoulder. The fellow was great, no doubt about it, and I felt like a fool asking him questions about matters outside the bounds of duty. "You're okay, Split. You could violate a hundred codes, as far as I'm concerned, and I'd swear before any court in the world that you're tops. But we've still got a problem with this tribe—and this girl."

"I'm not asking for compliments," Split said. "For the record I'm telling you what **did** happen, and what didn't. And here's what did." Now it was his turn to pace twice around the bench. "How do I begin?"

"With Omosla."

"Omosla comes to me often. She brings me food and drink. She hangs around like a pet. She doesn't touch me—anymore. I put a stop to that soon after the first time she put her arms around me. Yes, she did that. I was busy watching the sponge-trees move down the valley. She was nearby, murmuring words, most of which I could only half understand. I didn't stop her when she slipped her arms around me—not for quite awhile. I remember plenty well the way those pins in her elbow furs scratched my arms. They stuck in like thorns. Look, you can still see the marks." He rolled up his sleeves to show me the slight scars on his upper arms, just above the elbows. "I figured either she didn't know those pins were sticking me, or else it was some sort of tricky test that girls use on men to test their metal. So I took it, and didn't wince. Sure, I was enjoying letting her hug me. But after that one time I always kept my distance. This all happened when we first came. You'd think she'd have forgotten. Especially if she had a real husband somewhere on the scene."

I groaned. "Every tribe has strange customs. When the baby comes, that's when they'll insist on a husband."

"I wonder who it really is."

"Unfortunately we can't prove anything by giving the baby a blood test. These primitives wouldn't understand."

"Proofs are out," Campbell said.

"However, we still have the eyelash test," I suggested.

"You mean—"

"I mean that you and I are the only two human animals on this planet with eyebrows and eyelashes. When Omosla's baby arrives without a trace of an eyelash, that might go a long way toward convincing—"

"You'll help me fight it, then?"

"If you're sure you don't want to change your mind, throw out the Code, and claim the girl."

A look of disdain was all the answer Campbell gave me, at first. Finally he said, "You'd had ample reasons for nicknaming me Split, Captain. But so far, I've given you no grounds for applying the term to my personality. I prefer to remain a member of EGGWE, in good standing, and to return to Earth with a clear record. Let Omosla name the true father, whoever he is."

CHAPTER 8

The whole Benzendella tribe made its way across to the Kao-Wagwattl with only one casualty reported. Leeger, the short, slight guard who had once been brutally knocked out by Gravgak, was reported missing.

Everyone else came through without a scratch. It was a triumph for old Tomboldo. His superhuman courage had carried the day. Children were delighted over the adventure. Old folks were happy over achieving what they had feared would be an impossible undertaking. They could believe, now, that they would live all through to the end of the journey—for Kao-Wagwattl, the serpent river, was a legendary giver of life.

Campbell did not come. That was according to plan. He kept in touch with me by radio through the final hours of the twenty-mile crossing. "... Do you read me, Captain? I've drawn them to the north with fire bombs from the ship's guns.... They've never guessed your course."

"No signs of Gravgak? Or Leeger?"

"Not a sign. The city's empty."

"Keep on the radio, Campbell."

"Right, Captain. By the way, how is Omosla?"

"Expecting. I'll let you know. She still talks about the bravest man on the planet, someone named Campbell."

"H-m-m. You'll sort of look after her, won't you?"

It was two hours before dawn when the last of the tribe (Leeger excepted) gathered at the mountainside station to board Kao-Wagwattl. We waited for daylight. Strange smells filled our nostrils. Smells of wood fires, sparked to life by friction under the pressures of the crawling monster. Smells of rocks being ground to powder. Smells of the saccharine-sweet breathing from the pores of the thing itself, the giant Kao-Wagwattl.

The faint gray of dawn gradually changed to pink. In the growing light we could make out the contour of the vast misty creeping form. Its rounded sides moved along only yards from where we stood. As the light of morning came on we could distinguish the immense box-shaped scales that covered its sides. Clouds of sponge-trees rose and fell around it. Unrooted vegetation would sift downward, to be bumped into the air again, or to be rolled under. Small fires were continually being ignited by friction, and often smothered before they were well started. Sometimes the burning would creep up around the curved sides, only to be snuffed out by the surface-breathing of the massive thing.

I was relieved to note that the curved top—the "spine", so to speak—was so gradually rounded that there could be no danger of anyone's falling off. Its immensity had to be seen to be appreciated.

As to its length, I took the word of Tomboldo and others. It was endless. It wound around the whole planet like a fifty-thousand mile serpent that had swallowed its own tail. An unbroken rope of life, forever crawling.

A gigantic creature? A gargantuan vine? A living thing! I should not say that it was more animal than plant. When I asked Tomboldo's counsellors, Was it animal or vegetable, their answer was, Yes. Yes, **what**? Yes, it was animal or vegetable. They stressed the OR. Must it be one and not the other? Evidently the Kao-Wagwattl was not to be compared, not to be classified, but to be accepted—and utilized.

For this wandering tribe it was a means of escape from enemies, and a mode of travel. With the coming of daylight, they went to work.

Crude cranes. Swinging baskets. Hoists. One group after another was tossed up into the rubbery purplish-gray scales that covered the Kao-Wagwattl's spine.

No one cried out. The landing was soft. And harmless. The speed of the crawl was not great. It must have averaged not more than ten or fifteen miles an hour. But there were variations, to be taken advantage of. The outsides of a curve moved swiftly. Foresighted Tomboldo had selected the inside of a curve for our mounting, where the movement was sluggish. Younger members could leap across from an overhanging platform. Once safely in the folds of the surface, they could climb the rounded wall at their leisure.

Three or four hours were required for the entire tribe to get aboard. This meant that a long line was formed. Over a span of many miles this headless, tailless serpent became inhabited with tiny human fleas, figuratively speaking.

Among the stragglers who boarded last were a few older persons who had to be coaxed and pampered before they would get into the swinging basket.

Then, too, there was Omosla, looking very pretty and thoroughly frightened. She caused a slight delay at the very last by deciding it was time for her to have her baby.

CHAPTER 9

Finally we were all aboard, and the mighty Kao-Wagwattl, unaffected by this addition of a few specks of human dust, moved on at its dogged pace through the mountain valleys.

No lives had been lost. No one had been seriously injured. Tomboldo was the heroic leader. I went forward over the lumpy slabs of scales, to find him and congratulate him. He said, "The glad feelings are to be shared," and he spoke with high praise of my own help and that of my friend Campbell. "But we are not yet out of danger. Pass the word."

Pass the word. Keep down. Out of sight. For several days we would be crawling through the lands of savages.

Vauna found me. She had made sure that Omosla and the baby would have the best of care, and now she meant to look after me. "My dear one," she called me.

"Here, my dear one. I have your valuable coat. Come out of sight. The enemy must not see you."

I glanced up the long curved spine of Kao, moving steadily through the sunshine. Little groups of Benzendellas could be seen ahead, as far as the eye could reach. The young children of the party had never had such a trip before, and the older ones found it a strenuous game to keep them down out of sight. Following Tomboldo's order, they rapidly ducked down into hiding. The great rubber-like scales resembled up-ended boxes, set in criss-cross rows. The deep flexible crevices thus formed were ideal for hiding.

I needed my radio. I must talk with Campbell. Vauna had taken my coat.

She called to me. "Come, my dear one." She slipped down into a crevice a little to one side of the crest. "Come, I hear the voice of your friend Campbell in the box."

"I'm coming. Speak to him, Vauna. Tell him to wait."

"Shall I tell him the news?"

I didn't answer. The vertical surfaces of the scales folded together, parted, folded again, with the motions of the great creature, and for a moment I lost sight of Vauna. But I could hear her voice as I fought my way down to her hiding place. She was talking through the radio with Campbell.

"You are safe on the big silver ship?... Yes, we are on Kao-Wagwattl. I have been looking after Omosla...."

I could hear the eagerness in Campbell's voice as he asked about Omosla. Vauna answered him in accents of joy. "She has had her baby ... A little girl! Very beautiful. Already she looks like you. **She has precious little lines of hair on her eyelids, and above her eyes, just like yours.**"

The damage was done! There was no point in my lying to Campbell to spare his feelings. Her words were the simple innocent truth. She was happy and proud to tell the wonderful news. Her words implied that Campbell would of course come and join us when his work was done, so he could be Omosla's husband, as all the Benzendellas expected.

About all I could say to Campbell was, "What she says is true, Split. It's a beautiful baby. Any father should be proud. I have nothing to add."

For hours afterward I could think of nothing else. I sat hidden among the deep soft scales, listening. Now and then the gentle movement would cause the crevices around me to gape open, wide enough to reveal a strip of sky. I wondered if sometime I might catch sight of a space ship bolting off into the blue. The only sounds I heard were the faint muffled rumblings of the Kao-Wagwattl moving along, like gentle thunder echoing up from somewhere down in the earth. It lulled me into relaxation, yet I could not dispel the mental image of Campbell sitting there in the ship, alone, brooding over the news. And tempted, no doubt, to touch the controls and leave this planet behind him.

Later I talked with him again, but we did not mention Omosla. He said he was busy with his scientific findings. I relayed to him descriptions of the Kao-Wagwattl—the "inside" story, from one who was concealed within its scales. We were back to our original assignment, now. For days and days to come, we pursued the scientific facts, comparing notes by radio.

At air-cruise speed, Campbell made trips around the planet, and completed his charts and maps. He reported that the beautiful land toward which we were moving was indeed a land of promise. But he gave slower estimates of the Kao-Wagwattl's speed, and he estimated that it would take us the larger part of a year to reach our destination. However, he managed to get an inside view of the larger Benzendella tribes who dwelt there. They were truly waiting for old Tomboldo's return, and were firm in their faith that the rope of life, Kao-Wagwattl, would bring him.

Such were the scientific and ethnological studies that Campbell and I were to share, by radio, in the weeks and months to come....

Now Vauna was beside me. We, like the others, were settled down for the long journey.

Innocent Vauna! She was trying so hard to please me. She sat very close, whispering to me.

I listened, and smiled, and tried to take my thoughts away from the image of Campbell, his honor shattered by her recent words to him about the baby—a baby with eyelashes—a baby that resembled him.

If I remained silent, Vauna would tease me into talking with her. "Do my words displease you, Captain?"

"Your words please me very much."

"You do not look at me. You only look away. Do you want me to sit close beside you?"

I drew her in my arms and held her. In silence I thought a thousand thoughts that I had brought with me across millions of miles of space.

Later I said to her, "Your arms are warm. Why don't you take these fur things off your elbows, to be more comfortable?"

She smiled, and kissed me as I had taught her to kiss. "You want me to?" And she removed the furry white elbow ornaments. It was very strange.... While we hovered close, she whispered to me of the secrets of life on this planet, unlike any other world I had known. And there were curious legends of Kao-Wagwattl, things she had carried in her heart to tell me if such a time as this should ever come.

As she talked, the pressure of the scale walls around us increased. The great Kao-Wagwattl was evidently moving through a dip, so that its upper surfaces were compressed. There was no lack of air for breathing, but the darkness and the pressure added strangeness to the sensation. The tightness of Vauna's arms against my own caused my head to spin. Perhaps it was the fever returning from my recent illness. My arms felt the stinging sensation of being penetrated by needles. My thoughts flicked back to something Split Campbell had once told me....

Later, when the Kao curved over a summit, and the patches of sunlight dashed in, I suggested that Vauna go forward to see about her father. She answered me with a curious smile. I snuggled deeper into the shade of the scales and slept. Hours later, when I awakened, she was again beside me.

CHAPTER 10

If Omosla's baby had been a boy, I believe that old Tomboldo would have named it for the highest honor in the Benzendella world. He was searching for a successor. Not among the grown-up warriors and counsellors. Among the infants. He sought a child favored by nature. Omosla was a beauty and a court favorite, even though she had been a servant. And Campbell, who was considered to be her mate, (though marriage had been delayed by circumstances) was of course a renowned hero. If the child had only been a boy!

I was kept busy reporting the reasons for Campbell's absence. He had stayed with our ship to guarantee Benzendella safety. Yes, it was true that he could fly through the air and catch up with us. But there were duties which kept him away.

My excuses wore thin. Vauna and her father begged me to tell him, over the radio, that Omosla was growing into a person of sorrow. The shadow of tragedy hovered over her.

I complied. I talked, by radio, with Campbell. He was in another part of the land, now, pursuing the purposes for which we had come. My mention of Omosla's plight aroused his defiance. He said he would rather be a deserter than serve a captain who did not accept his word. "For the last time, Captain Linden, I repeat that I am not the mate of Omosla. Do you believe me?"

"I don't know what to believe," I said.

His radio clicked off.

Vauna and her father and I secluded ourselves among the scales and talked. My one question was, Could there have been any other person among them who had come from another planet?

"You and Campbell. No others."

"How can you be sure?" I pursued. "Suppose someone from my world wished to pass for a native. Suppose he should pluck the hairs from his eyelids and cut away his eyebrows. Would you know him to be an outsider?"

"Come," Vauna said. "We'll walk from one end of the tribe to the other."

While the great endless Kao-Wagwattl carried us on, through deep valleys and across wide plains, Vauna and I went about, day by day, studying the looks of each male member of the tribe.

I scrutinized the eyes of each. I listened to the native enunciations. I got acquainted with each man by name and personality. Vauna's friendship to all was a help. Through her I began to gain a bond of affection for all these people, deep and solid. Their ways became natural to me. In the night their sleep-singing could be heard, welling up softly through the scales within which they rested. In the mornings one could see the parties of agile ones gathering food and liquid fruits that rolled within reach along the sides of the moving Kao.

We crossed a series of islands. For long spaces there would be danger of dips under the surfaces of waters. We would close ourselves tightly within the waterproof interstices until the danger had passed. Later, when the slimy surfaces of the scales had dried off, we would emerge.

And now, out of a chance conversation, I learned of another danger which had been with us all along. Gravgak was also on the Kao-Wagwattl.

"How did you know this?" I asked Vauna sharply.

"Didn't my father tell you? I received a warning soon after we began the journey."

"Warning—from whom?"

"From Leeger."

"Leeger! I thought he was missing."

"He reappeared. He had known of our plan. He had boarded, somewhere. He was back there, beyond the end of our party. He shouted the warning to me. That is why you and I moved up the line, and have kept ourselves hidden."

"He shouted a warning to you—"

"That Gravgak is also on board, looking for me."

11.

Weeks earlier, a search party had given up. It had all happened quietly. Tomboldo had kept a few of his top scouts on the job (as I now learned) and for months after our journey had begun they had scoured the scaly surfaces of Kao-Wagwattl, looking in vain for Gravgak.

Could we rest assured, then, that Gravgak had been bluffed out? That he had given up his purpose of trying to take Vauna? That he had long since climbed off the Kao-Wagwattl and gone back home?

We hoped so. Nevertheless we moved cautiously as our searches took us back through the long line of Benzendellas.

Then, without warning, we suddenly came upon Leeger. He saw us from a distance of fifty yards or less. We had come to the end of our tribe's settlement—evidently beyond the end; for in the last quarter of a mile we had found no persons dwelling among the scales.

"He motioned to us," Vauna said. "I'm sure it was Leeger."

But Leeger had disappeared from view. Back of us now was the wilderness of scales, their curved surface glistening and alive with color as the endless crawling spine followed us out of the distant blue haze. Miles of Kao-Wagwattl, and nothing showing on the surface.

We were down, now, almost out of sight, yet peering over. Suddenly the form of Leeger bobbed up again, only a few feet from us.

"Go back!" Leeger cried, flinging a hand at us. "Go back! He's coming!"

It all happened in less time than it can be told. Leeger rose up to warn us. We saw the knife fly through the air at him. He fell with the blade through his throat.

On the instant we saw the dark muscular form of Gravgak rearing up among the scales. The green-and-black diamond-shaped markings on his arms and legs glinted in the light. He had hurled his knife true. Triumph shone in his murderous eyes. He had killed the man who had stalked him to protect Vauna and Tomboldo. And now he must have believed that one of his prizes was within easy reach.

His arm flashed upward. It held one of those rockstrung clubs that the savages used so skillfully.

The weighted club whizzed through the air. I swung Vauna off her feet. I'll swear the rolling movement of Kao-Wagwattl helped me or I wouldn't have succeeded. We tumbled into the crevice.

Then I scrambled upward. Another glimpse of Gravgak. He dived down among the crevices, moving in our direction. A moment of darkness.

The scale-tops closed out the light. When they opened, he was there, coming at us.

I locked with him. We fought. The movement of the surfaces gave us an upward thrust. I kicked and tumbled to the surface. He caught my wrist, but the upthrust of the Kao favored me and I jerked him upward, onto the top of the scales.

We fought in the open. The rubbery footing was deadly, but it played no favorites. I struck a heavy blow that made the green-and-black lined arms shudder. Gravgak's eyes flashed as he plunged back at me. I struck him again, with the full force of my body. He bounced and tumbled. He rolled out of sight. But not for long. It was an intentional trick. He disappeared in the crevice where Leeger had fallen. When he came up, the bloody knife was in his hand. I heard Vauna's warning cry.

I leaped down into the crevice. She was trying to get my coat. She knew there were explosives in it, if she could only get them into my hands.

No time for that. Gravgak leaped down at me. The knife was rigid from his hand, coming down with a plunge. I kicked back, floundering against the tricky walls of the scales, and Gravgak fell down deep where I had been. I saw it happen. A sight I never expect to see repeated.

His descent to the base of the scales, where the walls joined, might have been a harmless fall. Yet who knows how sensitive is the material of the vast living thing called Kao-Wagwattl? The knife plunged into deep **Kao flesh** beneath our feet. The flesh opened. Gravgak whirled, tried to escape the opening. His arm twisted under him. And went down. As if something drew it. His back—his whole body, from hips to shoulders—was caught in the gaping hole that he had seemingly opened with a plunge of the knife blade. It closed on him. It severed him. Part of him was gone. Before our eyes there remained his legs, cut clean away. And his head, and part of one shoulder.

The rest of him? It would not return to sight. Kao-Wagwattl was a living thing. When it wished it could devour.

Many of the tribe came back to this spot to examine what remained of the traitorous guard. I too observed him closely. I examined his eyes with a glass. Also the eyes of the murdered Leeger. Neither showed any traces of eyelashes or eyebrows.

12.

The tribe rode on tranquilly. There would be new legends of Kao-Wagwattl, after what had occurred. Many were the stories, and I relayed them to Campbell, at the ship, who faithfully recorded them all.

There was a tragedy to be added. It could not have been otherwise. For some months the news of Omosla and her little daughter had been vague. It was the Benzendella tradition that weddings should not be delayed for long after the arrival of the first-born child. It was rumored that this young mother now faced the shame of having been left without a mate. It was hard to get exact information. Even though Vauna and I had always sought an understanding between us, some things were not talked about freely. Deepest, most important truths in new worlds are often the most elusive. Now I questioned Vauna closely, and I learned of the tragic end of Omosla.

"She and her baby are no longer with us," Vauna said quietly. "It happened one night when the stars seemed very close. They say she had studied the sky each night, wondering which of the worlds beyond was the world of Campbell."

"And then?"

"Two of her caretakers saw it happen, but they could not stop it. With the babe in her arms, she walked over the side of Kao-Wagwattl. And went down. Under."

Vauna went on to tell me that Tomboldo had urged silence about it. He would always believe that the girl had lost faith too soon—that Campbell might have come back when his work was done. Moreover, Tomboldo felt that it was important to the morale of the tribe that both Campbell and I be held in high esteem.

When Vauna finished telling me these things, she said she would ask me the questions she had been saving for many days. "Did you believe, Jim, that you would find some other person among us from your world?"

"I didn't know."

"If you had found such a person, what would you have believed then?"

"That he, and not Campbell, was the father of Omosla's child."

"And what," Vauna asked, "are you going to believe about us when our child is born?"

13.

We were around on the other side of the planet by now. I estimated that we had traveled more than seven thousand hours.

By this time many things had happened. So much that I doubted my ability to convey all the news to Campbell so that he would get a clear understanding. I had lain awake nights trying to formulate my message. If my words failed, I only hoped that my tone of voice would convey my appreciation. My appreciation of him. Of what he had gone through. Of what he must yet go through.

He talked with me quietly through the radio, and I could visualize him as if I were sitting beside him again in the space ship.

"Yes, Linden. Go on. I'm listening."

I told him of the death of Omosla and the child. He was deeply grieved. It was a long time before he found voice to speak.

"Go ahead, Linden. I'm listening."

"I have more news," I said. "But tell me of yourself, Campbell. Have you gone ahead, playing your lone hand?"

"I've found my way into the customs of the savages, Linden. They have their own legends of Kao-Wagwattl. I can predict that in time the gap can be bridged between them and the Benzendellas. If we work carefully—men like you, Linden, working from within, and other agents from EGGWE that are sure to follow. I believe this planet can be spared the torments of great wars."

"Yes, Campbell ... and you, personally ... are you well? Are you still bristling with your usual self-discipline?"

"In case you have any doubts about the matter," his voice was slightly caustic, "I haven't broken the Code."

"In Omosla's case I wish you had," I said.

"I wish it too," Campbell's voice came back, now in a lowered tone. "I loved Omosla. I would have been her mate, gladly."

"But you were, Campbell."

"Now, don't start that again, Linden, or I'll—"

"Wait, Campbell, don't cut me off. You must hear all of my news, first. Most important of all, old Tomboldo has chosen my own son to be his successor. He'll be groomed for the job all through his childhood, and I've decided to stay right here, Code or no Code, and see him through."

"Your **son**?" Campbell's voice was mostly breath. "Who are you talking about?"

"Our baby—Vauna's and mine. It's several days old. Doing fine. Has eyebrows just like mine. Chalk-dust skin like hers."

Campbell blurted. "Do you mean to tell me that as soon as you and Vauna boarded the Kao—"

"The ways of life on this planet are something you and I ought to know about, Campbell. Listen closely—"

"Shoot!"

In words of one syllable I explained, then, what I had at last learned: that the human beings of this planet were not precisely like those of the Earth. They were unquestionably related, somewhere back down through the ages. But Nature had worked a significant change in the process by which new life could be started. Fertilization in the female was accomplished by her own action and her own preference. Nature had equipped her arms—

"Arms, did you say?" Campbell fairly shouted through the radio. "Go on."

I continued. Nature had equipped her arms, I explained, with tiny thorn-like projections which could penetrate the arms or sides of the male like needles. By this means she drew blood from his bloodstream. A very slight transfusion of male blood into the female bloodstream was the act that accomplished fertilization.

"You see, Campbell, woman does not bear a child except by her own premeditated choice," I explained. "You and I were puzzled by the elbow furs all these women wear. Now you see. It's a natural bit of extra clothing. The dictates of modesty."

"Well!" Campbell said. "Then you and I allowed ourselves—"

"We were simply chosen. Not knowing the score, we were innocent bystanders—well, more or less innocent—and pitifully ignorant. Unfortunately for us, these were matters the Benzendellas don't talk about freely."

Campbell paused for a moment of confused thinking. "Just a minute, Captain. I've been observing these savages—home life and all. There's no lack of normal affections among them, in our own sense of the word. They're equipped physically, just as we are—plus the arm thorns. They have the same organs, the same functions—"

"For purposes of affection, yes. But the arms—that's separate—for conception."

"Well I'll be blasted!" Campbell was speechless for a long moment. Then, "I think I'll go back to Earth."

I was not surprised at his decision. It was what I expected, what I would have advised. He had had more than one man's share of this planet, for one who didn't expect to take root here. But my own life here was just beginning.

I had thought it out. My guess was that my long record of service for the EGGWE could withstand some variation. An application for release would very likely win an approval, especially in view of my change to serve the EGGWE purposes even better by becoming a Benzendella.

When I announced this plan, by radio, to the new Captain Campbell, formerly known as Split, but now commonly referred to on this planet as the hero of the Benzendella migration, he said he was not surprised. "Congratulations, Linden, for knowing what you wanted. Stay aboard that Kao-Wagwattl. There's a beautiful land waiting for you up ahead."